THIS BOOK BELONGS TO:

_____

Dedicated to my son.

All rights reserved.
No part of this book may be reproduced in any form or by any means, electronic or mechanical, and no photocopying or recording, unless you have written permission from the author.

ISBN 978-1-958985-17-5

Text copyright © 2024 by Mimi Jones

www.joeysavestheday.com

A Mimi Book

Welcome to the Wonderful World of Shoebill Storks

These tall and remarkable birds are closely related to pelicans and herons. Their distinctive appearance sets them apart as captivating creatures.

Pelican

Heron

Balaeniceps rex, commonly known as the Shoebill, is a unique bird species often referred to as the Shoebill stork.

The Shoebill stork can grow as tall as five feet.

The shoebill is a big bird that is found in the swamps of tropical East Africa.

Africa

In the wild, they have an average lifespan of up to 35 years.

The Shoebill is known for its large, shoe-shaped bill.

The Shoebill primarily wades through marshes and swamps.

It can fly short distances when necessary, using its powerful wings to stay above the water.

These distinctive birds are typically found in Uganda, South Sudan, Tanzania, and Zambia.

It is known for its distinct, prehistoric appearance.

Shoebills depend on their wetland habitats for survival, needing access to water for fishing and other vital activities.

Shoebills are well adapted to their habitat and are recognized for their distinctive shoebill-shaped bill, specifically designed to capture prey in the murky waters where they reside.

This bill, combined with their sharp eyesight, makes them formidable hunters.

Shoebills are powerful carnivorous birds that mainly consume fish, small crocodiles, snakes, lizards, and other prey.

They forage for their food by wading through the shallow waters of swamps and marshes. They use their distinct, shoe-shaped bills to catch and consume their prey.

Unique Trait:

Their beaks are impressively long, reaching up to 9 inches in length, ranking among the longest in the bird world.

8 FEET

The wingspans of Shoebill storks can reach up to 8 feet.

They have the potential to reach a weight of around 15 pounds.

15 LBS

The magnificent Shoebill stork is renowned for its ability to produce powerful and unique sounds that have been compared to the rapid fire of a machine gun or the rhythmic beating of tribal drums.

MAGNIFICENT!

5000 To 8000

The population of the shoebill stork is estimated to range between 5,000 to 8,000. This species is classified as "Vulnerable" due to habitat destruction, disturbance, and hunting.

UNIQUE

Shoebill storks are unique birds. Their distinctive appearance and remarkable hunting skills make them stand out in the avian world.

Story Time: A Swampy Adventure

In the heart of the African marshes, Billy the Shoebill Stork stood tall and still, his large, shoe-shaped bill ready for action. The early morning mist hung low over the water, creating an ethereal atmosphere. Billy's sharp eyes scanned the surface, searching for the slightest ripple that would signal the presence of prey. With a sudden, swift movement, he plunged his bill into the water and emerged with a wriggling fish, his first catch of the day.

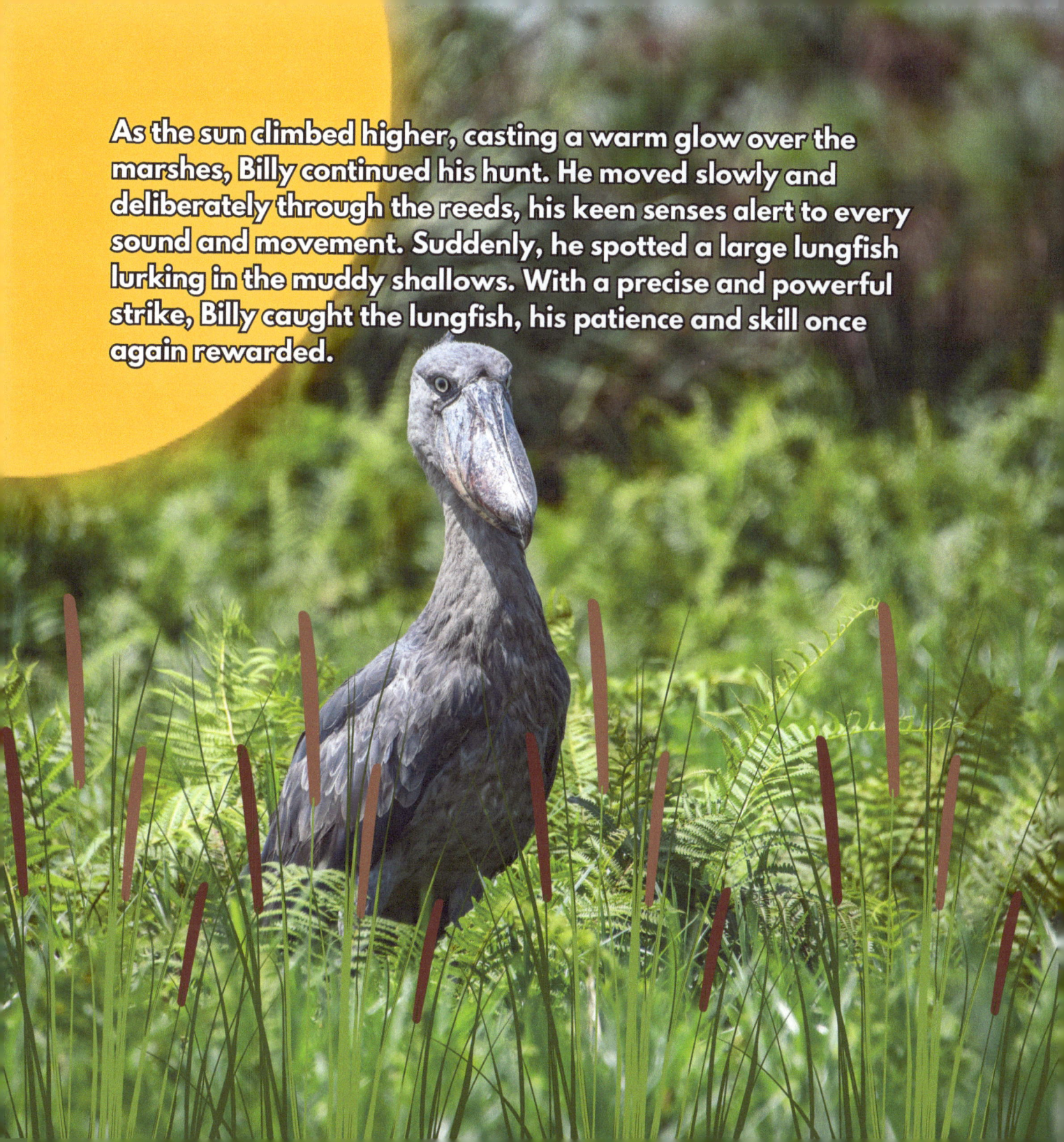

As the sun climbed higher, casting a warm glow over the marshes, Billy continued his hunt. He moved slowly and deliberately through the reeds, his keen senses alert to every sound and movement. Suddenly, he spotted a large lungfish lurking in the muddy shallows. With a precise and powerful strike, Billy caught the lungfish, his patience and skill once again rewarded.

Satisfied with his successful hunt, Billy returned to his favorite perch, a tall tree overlooking the marshes. He preened his feathers and settled in for a well-deserved rest. As the day turned to dusk, Billy reflected on his adventures, feeling a deep sense of contentment. The swampy marshes were his home, and he was proud to be a part of this vibrant, ever-changing world. With the promise of new adventures tomorrow, Billy the Shoebill Stork closed his eyes and drifted off to sleep.